Postmodern Physics

Principia Quantuma Mechanicum

physicis sit amet !

Greg Feild

November 1, 2020

'Second Edition'

About the author:

Greg Feild is a physicist with a PhD from the Pennsylvania State University.

He is the author of a quantum mechanical theory of everything,
and has written several books on the subject.

He is also the manservant to two very bad dogs.

> The deluded, imagining trivial things to
> be vital to life, follow their vain fancies and
> never attain the highest knowledge
>
> But the wise, knowing what is trivial and
> what is vital, set their thoughts on the supreme
> goal and attain the highest knowledge
>
> -- [the Buddha]
> *The Dhammapada*

Abstract:

In this book, we examine the world in the light of our new universal model of physics.

We will visit all of our favorite themes and continue the investigations into several
topics still incomplete since our last book "Toward a Metaphysics of Mass and Motion",
such as; our new *physical interpretation* of spin and our new *physical model*
of the only three elementary subatomic particles; the photon, the neutrino and the electron.

The universal model is field, and vacuum, free. Those paradigms are tired, and retired.

There is **only** mass in motion.   **There are only atoms and the void**

        For I see a man must either resolve to put out nothing new
        or to become a slave to defend it.

                                                -- Isaac Newton

Greek philosophy for physicists:

    And assuredly no one will argue that there is any other method [than dialectic] of comprehending by any regular process all true existence, or of ascertaining what each thing is in its own nature; for the arts in general are concerned with the desires and opinions of men, or are cultivated with a view to production and constructions; or for the preservation of such productions and constructions; and as to the mathematical sciences which, as we were saying, have some apprehension of true being -- geometry and the like -- they only dream about being, but never can they behold the waking reality so long as they leave the hypotheses which they use unexamined, and are unable to give an account of them.

For when a man knows not his own first principle, and when the conclusion and intermediate steps are also constructed out of he knows not what, how can he imagine that such a fabric of convention can ever become science?

-- Plato
Republic

    Now, there are certain philosophers who, as we have intimated, themselves both affirm that it is possible that the same thing may and may not be, and that they really think so. [!]

This principle, however, do many investigators of nature employ. But we just now have assumed it as a thing impossible, in the case of an entity, that it should be and not be at the same time; and by means of this have we demonstrated that this is the most firm of all first principles.

-- Aristotle
The Metaphysics

It is always good to remind ourselves of the facts, to remind ourselves of what we actually know.

This method [*conceptual* revision], I believe, is one of the ways to make progress in philosophy. When confronted with an intractable question such as is presented by a clash of convincing default positions, don't accept the question lying down. Get up and go behind the question to see what assumptions lie behind the alternatives the question presents. In this case [dualism vs. materialism], we did not answer the question in terms of the alternatives presented to us but we *overcame* the question.

The fact that there are no universally accepted procedures for solving philosophical problems does not mean that anything goes, that you can say anything or that there are no standards. On the contrary, precisely the absence of things as laboratory methods to fall back on forces the philosopher to even greater degrees of clarity, rigor, and precision. In philosophy there is no substitute for a combination of original, imaginative sensibility, on the one hand, and sheer intelligent, logical rigor on the other. The rigor without sensibility is empty, the sensibility without rigor is a lot of hot air.

-- *Mind, Language, and Society*
John R. Searle

The model drawn as a hypothesis, as a tentative tracing, acquires once drawn the appearance of permanence, solidity, and reality --- so much so that even researchers who begin with an awareness of the hypothetical nature of their models often end by believing in their reality.

All disciplines tend to reify their abstractions, to mistake their main conceptual terms for concrete things. From long residence in our minds, our most familiar terms --- no matter how abstract they are --- are likely to be seen closer to home and hence more concrete than whatever we're supposed to be investigating "through" them.

-- Words and Values
Peggy Rosenthal

"For the want of a nail … the kingdom was lost."

Postmodern physics:

One could argue that physics is currently in the death throes of its postmodern phase with its antediluvian creation myth and a collection of just so stories concerning non-locality and reverse causality.

All jokes, hoaxes, and irony aside, we shan't go there.

At least 'the they' of modern physics have eliminated the completely unacceptable and 'occult' action at a distance from quantum field theory! They also replaced the improbable notion of electron spin with a collection of totally impossible ghostly supporting structures to shore things up.

For want of spin … all physics was lost !

In our new model spin is real. Angular momentum is the fundamental unit of interaction and the defining characteristic of photons and leptons. All physics is cast in terms of angular momentum and angular variables.

As a result, we express Newton's laws in terms of angular momentum in the center of mass of an N body system. This formulation is valid not only in any inertial reference frame, but in all accelerating reference frames as well. (See the next section.)

In our model, particle spin and planetary spin, etc. are instances of absolute motion that cannot be denied, or 'transformed' away. We fix our inertial reference frame with respect to the fixed background of space.

The results of quantum field theory and general relativity (will) both 'reduce' to our new model.

In addition, although our model was constructed with hindsight, one can derive the rules of quantum mechanics from our physical theory of particle propagation and interaction.

One cannot derive quantum mechanics from the Copenhagen interpretation.

We will begin this book with several pages of 'summary material', followed by the introduction.

beware *the 'they'* !

Interlude: More Epigraphs

On the Method of Mathematics:

    The Euclidean method of demonstration has brought forth from its own womb its most striking parody and caricature in the famous controversy over the theory of *parallels*, and in the attempts, repeated every year, to prove the eleventh axiom. This axiom asserts, and that indeed through the indirect criterion of a third intersecting line, that two lines inclined to each other (for this is the precise meaning of "less than two right angles"), if produced far enough, must meet. Now this truth is supposed to be too complicated to pass as self-evident, and therefore needs a proof; but no such proof can be produced, just because there is nothing more immediate. . . .
In fact, it seems to me that the logical method is in this way reduced to an absurdity. . . .
But that axiom is a synthetic proposition *a priori*, and as such has the guarantee of pure, not empirical, perception; this perception is just as immediate and certain as the principle of contradiction itself, from which all proofs originally derive their certainty.

On Man's Need for Metaphysics:

    We also find *physics*, in the widest sense of the word, concerned with the explanation of phenomena in the world; but it lies already in the nature of the explanations themselves that they cannot be sufficient. *Physics* is unable to stand on its own feet, but needs a *metaphysics* on which to support itself, whatever fine airs it may assume toward the latter. For it explains phenomena by something still more unknown than are they, namely by laws of nature resting on forces of nature . . .

                                                        --- Arthur Schopenhauer
                                                        The World as Will and Representation

Newton's laws: (From "Revenge of the Sinister Universe")

In this section we present a synopsis of the 'universal' formulation, or expression, of Newton's laws, as proposed in our last book, "On Rotation".
[This formulation is good in *any* reference frame.]

Newton's first law:

If a particle does not experience a change in *angular momentum* relative to *any* arbitrarily chosen point in the universal reference frame, then the particle is considered to be free (i.e. there are no net forces acting on it).

For a two body system, if there is no change in the angular momentum of *either body* comprising the two body system relative to *any* arbitrary point (this 'excludes' the choice of points lying along the unit vector, **r**), then the particles are *not interacting*.

Newton's second law:

A particle that undergoes a change in angular momentum relative to our arbitrarily chosen point (i.e. the origin of our coordinate system) is said to experience a net torque, τ ;

$$\tau = d\mathbf{L}/dt = \mathbf{r} \times \mathbf{F} \tag{a}$$

where the force, **F**, is defined by

$$\mathbf{F} = d\mathbf{p}/dt = d(m\mathbf{v})/dt = m\, d\mathbf{v}/dt + \mathbf{v}\, dm/dt \tag{b}$$

For two body 'central force' motion, the torques experienced by each individual body relative to our chosen reference point, are equal and opposite;

$$\mathbf{r}_1 \times \mathbf{F}_1 = -\mathbf{r}_2 \times \mathbf{F}_2 \quad ; \quad |\mathbf{F}| = |\mathbf{F}_1 - \mathbf{F}_2| \tag{c}$$

⇔

Newton's third law:

During a two body interaction, the two bodies will undergo equal and opposite changes in their respective '*actions*'; i.e. they will have equal, and 'opposite', changes in kinetic energy.

$$\delta \int d\mathbf{L}/dt \cdot \omega \, dt = 0 \tag{d}$$

where

$$\tau_{TOTAL} = \tau_{FORCE} + \tau_{SPIN} \tag{e}$$

and

$$\tau_{SPIN} = \mathbf{l}_1 \times \mathbf{B}_2 + \mathbf{l}_2 \times \mathbf{B}_1 \tag{f}$$

Newton's universal law of gravitation, expressed in the center of mass of a two body system, becomes; [note: equation (h) has been corrected relative to previous books.]

$$F/E_{TOT} = K^*(c/R)^2 \mu - K^*(\mu v^2/R^2) - K^*(c^2 l^2/\mu R^3) \tag{g}$$

$$K = (G/c^2) \tag{h}$$

where the second term on the right hand side of equation (g) is the *coriolis* force; our answer to spacetime disturbances. [This extra term will describe the rotation of galaxies without the inclusion of 'dark matter'; and the perihelion of the planet Mercury.]
[For galaxy rotation, one *must* include the 'relativistic' mass of the individual bodies due to spin.]

Since our new coriolis force term goes as $1/R^2$, the classical and quantum mechanical conservation of the first three integrals of the motion, E, L, L_z, is still guaranteed.

Equation (d) represents the classical 'minimization of the action' expressed in our new angular variables. Relativistically, all interactions minimize the change in kinetic energy and equation (d) becomes

$$\delta \int (m(t) - m_0) \, dt = 0 \tag{i}$$

In the final formulation, equation (g) will also have to include a spin term.

Particle [electron] in a box: (From "On wave particle duality …")

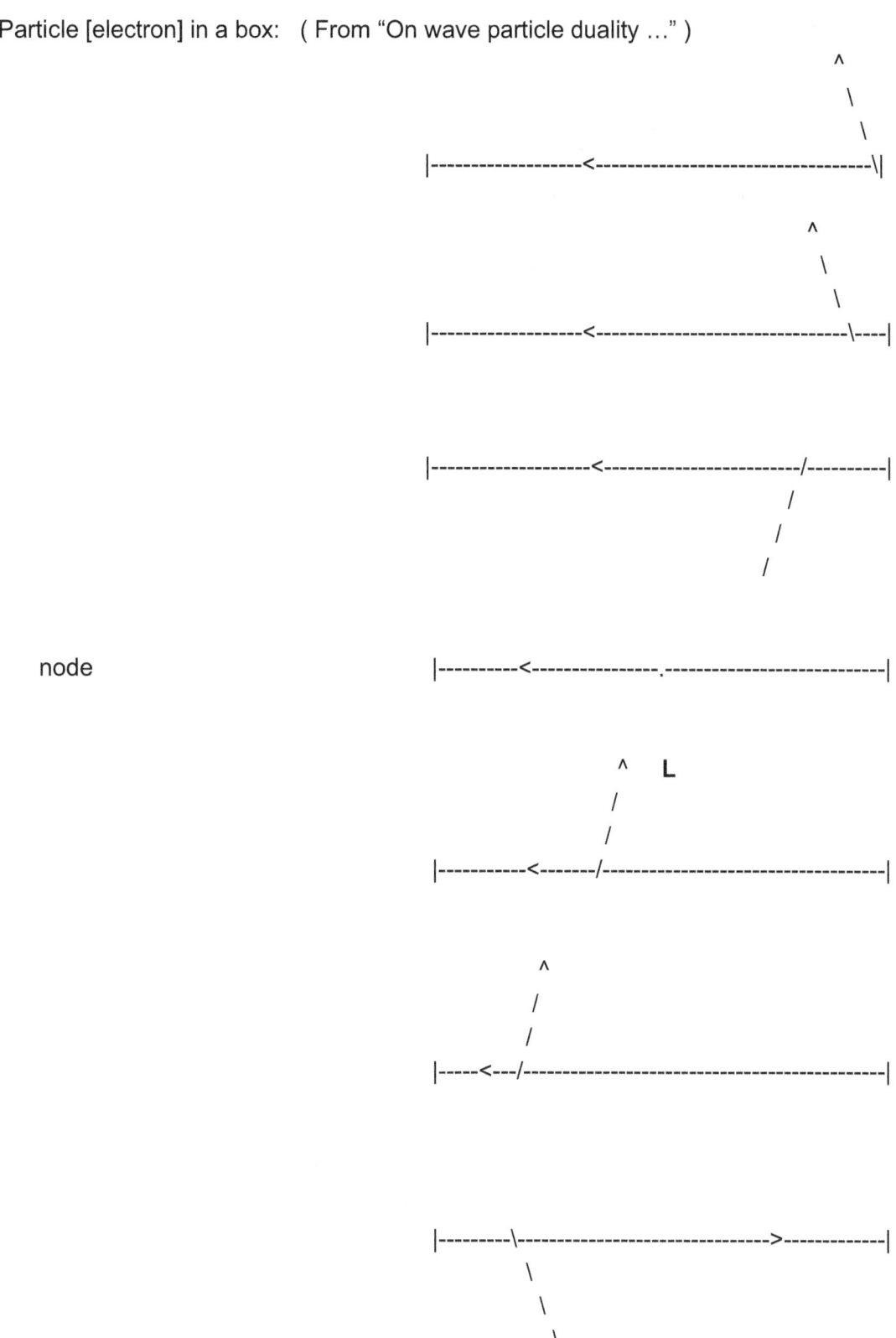

Figure A; At the 'node', the angular momentum vector **L** points up out of the page.

The neutrino: (From "On Rotation")

We imagine the neutrino to be one quantum of action per unit volume of space, resulting in an available, or 'functional', angular momentum of $\hbar/2$.

The neutrino is a spinning point particle with three angular momentum vectors.

Only one component of the angular momentum may be projected along the direction of travel, and the total angular momentum 'gimbles' about this direction, as do the other two components. [The average of the other two components will obviously be zero!]

The gimbling of the neutrino [a spinor] is illustrated in Figure B.

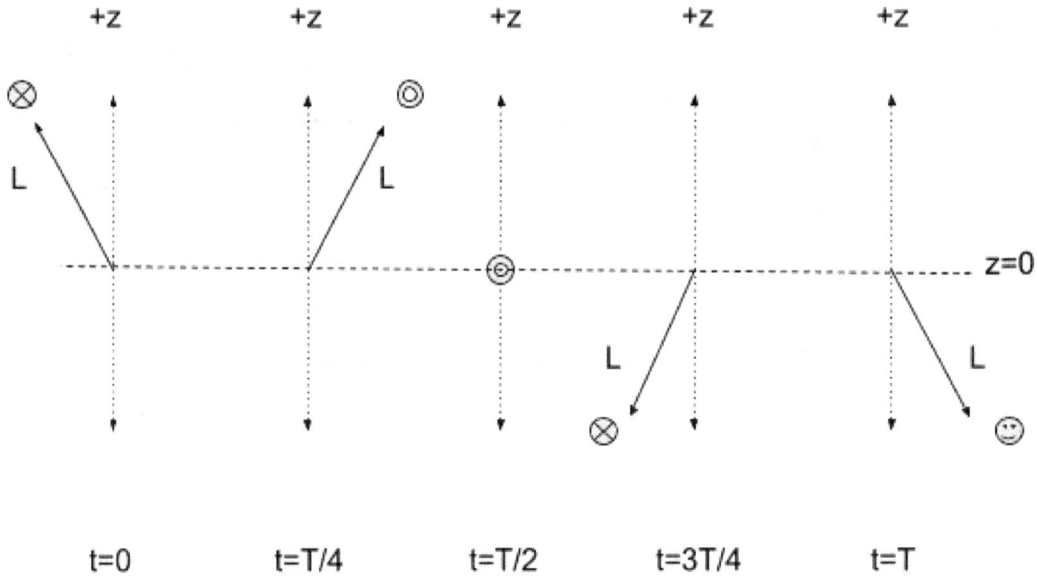

Figure B: At rest with a lepton traveling in the z-direction. The spin angular momentum vector 'precesses' about the direction of motion, tracing out a closed, three dimensional figure eight. The x symbol represents motion into the page. The dot symbol represents motion out of the page. At time T/2, we see the angular momentum is *perpendicular* to the direction of travel. (This is when the lepton engages in 'virtual' interactions.)

To represent an antilepton, simply swap the x symbols and the dot symbols.

The photon: (From "On Rotation")

The photon is one unit of angular momentum, *the fundamental unit of angular momentum*, Planck's quantum of action; h. The angular momentum of the photon is always h, and the speed is always c. So, how does the photon gain and lose energy and linear momentum during an interaction?

In our model, the projection of the photon angular momentum vector, along the direction of travel, varies sinusoidally at the frequency that defines the energy and momentum of the photon according to the usual relations;

$E = h\nu$
$p = h\nu/c$ (j)
$m = h\nu/c^2$

Let us say the photons 'gyres' as illustrated in Figure C.

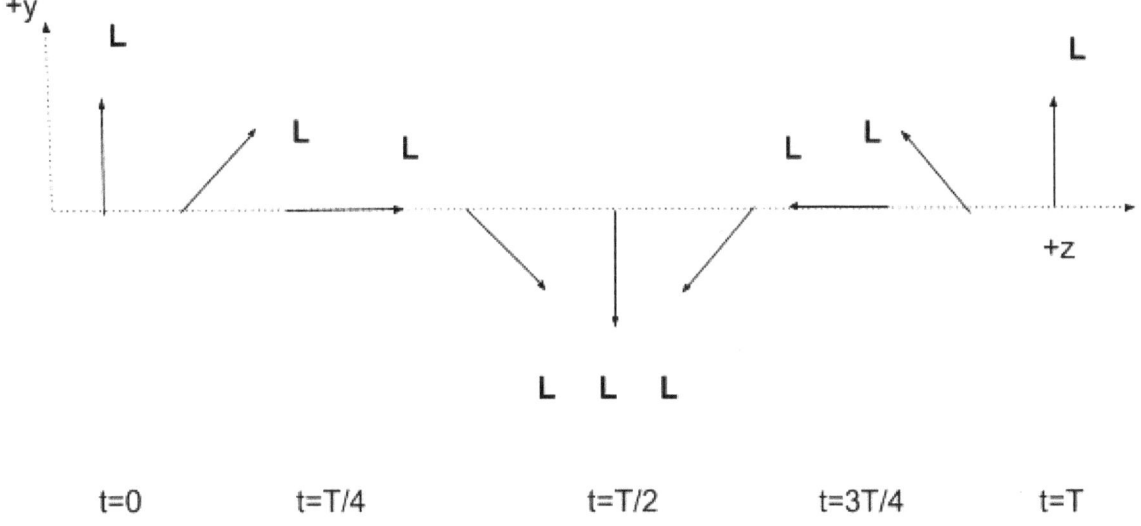

t=0   t=T/4   t=T/2   t=3T/4   t=T

Figure C: The photon rolls, or 'gyres', along the direction of travel projecting the spin angular momentum vector sinusoidally along the direction of propagation while maintaining a constant polarization and angular momentum; L = h. We could also say that the photon '**spirals**'. At t=0, the plane of the photon spin is *coplanar* with the z-axis.

A universal metaphysics ?: (From "On Epistemology and Ontology")

The fundamental unit of matter is the neutrino.
The neutrino is a gravitational-point-mass-charge;
or, a point, gravitational, magnetic moment, μ

$$\mu \ = \ h^{bar}/2c \ (\ 1 + \tfrac{1}{2} v^2/c^2 + \ldots) \tag{k}$$

The elemental unit of matter is the dipole! This may spark thoughts of 'dialectic' in our philosophical readers; however, the magnetic dipole is represented, mathematically and schematically, as <u>closed loops</u> of 'magnetic field lines' ! This picture of the neutrino is illustrated in Figure D.

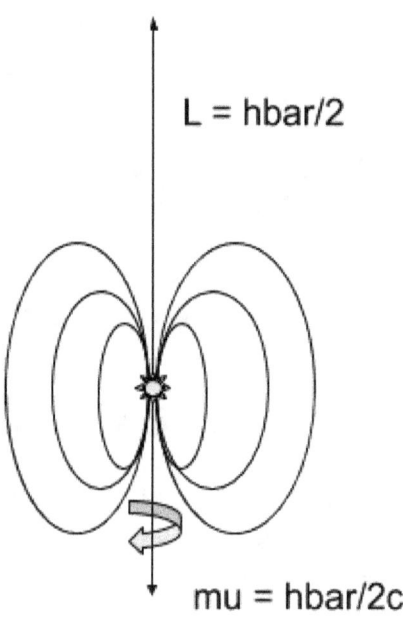

L = hbar/2

mu = hbar/2c

Figure D: The angular momentum, magnetic moment, and magnetic field lines of
a point, gravitational, mass dipole; aka the neutrino!

The neutrino is spinning to the left, with the canonical definitions for the directions of L and μ given by the standard model.

Thus, there are no magnetic monopoles !

The electromagnetic charge: (From "Revenge")

Interactions conserve spin, mass, and charge. In the universal model, we have generalized the concepts of electric charge and mass, to include all conserved quantities.

We define the electromagnetic charge to be

$$Q_{EM} = e \hbar / 2c \; \mathbf{s} \qquad (l)$$

and the mass charge to be

$$Q_{MASS} = m \hbar / 2c \qquad (m)$$

These quantities must be conserved at every 'vertex'.

The nature of these new charges are summarized in Table A.

Force	Coupling constant	Conserved current	Rotation basis	Conserved charge
Electricity	e/m_e	Mass isospin	e, mu, tau	$e*\hbar/2*c \; \mathbf{s}$
Gravity	m_e/e	Charge isospin	e, ν_e	$m*\hbar/2*c$

TABLE A: Table of coupling constants, conserved currents, and charges: Note: $m_e/e = m_\nu$

The leptonic table: (From "Revenge")

LEPTONS   ANTI-LEPTONS

electron	electron neutrino	PARITY ⇔	electron antineutrino	positron
⇐	CHARGE	MASS ⇕	CHARGE	⇒
muon	muon neutrino	PARITY ⇔	muon antineutrino	anti-muon
⇐	CHARGE	MASS ⇕	CHARGE	⇒
tau	tau neutrino	PARITY ⇔	tau antineutrino	anti-tau
⇔	mass isospin	charge isospin ⇕	mass isospin	⇔

TABLE B: The leptons and their interrelations; or the kleptogenesis of the leptoquarks.

Any lepton can be 'generated' from any other by the appropriate applications of the parity operator, the mass isospin operator, and our newly proposed 'charge isospin' operator.

Using various combinations of the step up and step down operators of SU(2) [charge isospin] and SU(3) [mass isospin], plus the parity operator, we can write any quantum mechanical interaction current in terms of the 'fundamental' neutrino neutral current.

Units and dimensions:

It is time to take the plunge and redefine the electric charge in terms of mass. This is the last necessary step to put gravity and electromagnetism on an equal footing. In our model, the fundamental coupling charge for all interaction is the particle mass. The charge to rest mass ratio of the electron, e/m_e, is now a coupling constant that acts to distinguish the electron from the neutrino. The new 'electric charge' is $m(e/m_e)$. We can see that the constant e must now have units of mass to make the new units work out correctly.

We believe that the phenomenon of electromagnetic induction is responsible for the increase in electron mass relative to the neutrino and endows it with an electric charge and thus an additional electric field. It is the motion of mass, rather than electric charge, that induces magnetic forces in our new model leading to the hypothesized magnetic moment of the neutrino. Obviously, this will involve the redefinition of several units, making them all Coulomb free. Mass currents and electrical currents will then be on equal footing, and will interact with one another, but with very different coupling strengths, of course.

We have yet to work out the details of what these new units might be called and how they will affect other units and their designations.

Details ! See "On Rotation" for a few more thoughts and details on this subject.

Errata:

We have conjectured (1) that the value or magnitude of e is the ratio of the mass of the electron to the mass of the neutrino

$$e = m_e/m_\upsilon \tag{o}$$

We have decided this is (obviously!) wrong as it does not work out mathematically or jibe with our new redefinition of the units of e as described above. Why I have held onto this idea so long although I could see there was something wrong with it is a mystery to me.

sigh

Introduction:

The universe is governed by one force; gravity or electromagnetism, call it what you will.

There are two fundamental matter particles; the neutrino and the electron.

The difference between the strengths of gravity and electromagnetism, and the difference between the mass of the neutrino and the mass of the electron, are, both, due to the same fundamental phenomenon; electromagnetic induction.

Electromagnetic induction is the 'symmetry breaking mechanism' of the universal model.

Electromagnetic induction is also the most 'difficult' concept in our model, and is totally derived from the classical idea. We take it to be an empirical discovery, and thus 'irreducible' theoretically.

The earth is a terrible place to do physics. It is round and it spins. Locally, it is flat and full of friction; air resistance, viscous fluid flow, sticky surfaces, rough terrain, etc.

We have found during our investigations that we must agree with Aristotle, and many other ancient Greeks: the natural state of a particle is to be 'at rest', and all particles that are <u>in motion</u> *tend toward* a state of rest even though nature is nothing more than particles in motion, and/or constant change and becoming! More on this later !

In our model, the <u>total force</u> between any two objects (electrons, planets, etc.), normalized by the total energy of the system, and cast in planar coordinates is

$$\mathbf{F}/E_{TOT} = K*(c/R)^2 \mu - K*(\mu v^2/R^2) - K*(l^2/\mu R^3) \qquad (1)$$

where μ is the reduced (relativistic) mass, and

$$K == (G/c^2 - (e/m_e)^2(\mu_0/4\pi)) \qquad (2)$$

Equation (1) is essentially all of physics in a nutshell !   :)

That is the premise of the universal model.

Symmetry:

The symmetries of the universal model are; up, down, left and right --- essentially, the conservations of parity and helicity. The relationship between these quantities is conserved under rotations. The invariance of any system under rotations insures the conservation of angular momentum and kinetic energy; the only two *truly fundamental physical quantities*.

The invariance of systems under linear transformations insure, or demonstrate, the conservation of linear momentum *and* the conservation of energy. This is so because in our model the position and momentum coordinates are variables, while *time* is merely a parameter. Time is essentially over *constrained* and can not vary independently relative to the space coordinates. In addition, the concept of potential energy has been 'discredited' in a technical sense, although perhaps not operationally. Hence there is no need for a time translational invariance to demonstrate the conservation of energy. ALL energy is kinetic energy.

Due to their nature as *spinors*, fermions that spin to the left, spin to the left whether or not they are pointing up or down. They spin to the left whether or not the spin is projected along the direction of propagation or against it. Hence, one cannot change the helicity of a particle by making a Lorentz boost as is commonly believed.

Fermions that spin to the left are *matter,* and have a parity of +1. Fermions that spin to the right are *antimatter,* and have a parity of -1.

Protons *spin to the right.* Protons have a parity -1. One may call them antimatter, if it really matters! All this has been explained in our previous books, so we won't go into the details of this determination here.

The essentially random assignment of a parity of +1 to the proton in the standard model has led to all kinds of excitement and confusion concerning parity and CP violation.

A tempest in a teapot. Much ado about nothing !

As for *gauge symmetry,* it is just not a thing. As we have shown, mass, or mass-energy, is the fundamental coupling charge. The rest mass of a particle corresponds to the 'global charge' of the original electromagnetic gauge theory, and the relativistic mass is the 'local charge.' The universal wave function is then

$$\psi = \exp(i(p \cdot x - (m-m_0)c^2 t)/\hbar) \qquad (3)$$

and, the Dirac equation is

$$i\hbar\, \partial\psi/\partial t = -ic\hbar\, \alpha \cdot \nabla \psi \quad \longrightarrow \quad mc^2 \psi = c\alpha \cdot p\, \psi + m_e c^2 \psi \qquad (4)$$

Cause and effect:

The *inter*action between any two objects (or collection of objects) is due to the equal and opposite forces objects exert on each other as indicated in equation (1). All *inter*action is action and reaction. As Kant has taught us, the idea of causality is an *a priori*, or intuitive, conception imposed on events by consciousness or the mind. In more modern terms, we would say this way of perceiving the world is hardwired in the animal brain due to hundreds of thousands, if not millions, of years of evolution. The idea of causality arises from the temporal nature of these interactions. Time flows forward by definition. Don't let anyone tell you otherwise!

The confusion about cause and effect and the 'direction' of time began early in the history of physics and predates both quantum theory and relativity theory. One wants to describe the motion of a projectile in the earth's gravitational field. We know the projectile is going to travel *forward* in both space and time so we extract the appropriate physical parameters from our essentially metaphysical notions of space and time and construct the equations of motion.

Then, some Clever Trevor notes that one can put a minus sign in front of every t variable leaving the equations of motion 'unchanged' and thus concludes it is some great mystery why particles do not travel backwards in time. *There is no mystery.* The equations in question were specifically designed to describe forward motion. The equations are simply a tool designed by people to describe specific observations and physical situations. They have no meaning or inherent value or significance themselves. The idea that time flows forward was a preconception and condition in the construction of the equations. One can easily fix the mystery by defining time to be 'positive-definite.'

The real mystery is why *anyone* would find this interesting, confusing, exciting, or indicative of some deep conundrum concerning the physics of time.

The equal and opposite forces of action and reaction between two particles or 'objects' are due to the constant and continuous exchange of energy and momentum by way of the virtual photon connecting the two. The frequency of the virtual photon, and thus energy exchanged, is proportional to and varies with the distance between the objects. This variation in frequency, and thus the transmission of the varying force between the two, is 'instantaneous' and requires the specification of a common measure of time. This model is in stark contrast to modern field theories which assume that the propagation of internal forces occur at a finite speed. One object emits a virtual particle and the other absorbs it. But which is the source and which is the sink? The more massive object is always the source, of course! Modern field theory fails to recognize the fundamental difference between real and virtual photons. Only real photons are emitted and absorbed, and travel through space at a finite speed.

Time and space:

   Oh my gosh! It is going to take teams of psychologists, sociologists, historians, philosophers, and, yes, even physicists, many, many years to unravel and understand how people have arrived at their current incoherent and childish conceptions concerning time and space.

We certainly do not have the time, patience, or inclination to tackle the issue here. Physicists should stick to physics and leave matters on ontology, epistemology, and mathematics (see the disaster that is gauge theory!) to the respective professionals. In addition, mathematicians should not try their hand at physics. Of course, the pursuit of physics touches upon all these pursuits, particularly mathematics, but one must always start with a physical model. Do the physics, then the math, and then the philosophy. The true scandal is the manner in which philosophers have capitulated to the physicists and try to lend support and justification to all their pathetic ideas.

   Philosophy is no longer the handmaiden of physics. Those days are long gone and ended with the Age of Reason.

   As for time and space, one cannot speak of such things in terms of physics without respect to a given problem and choice of reference frame.

   Whether one is studying the interaction between two fermions, planets, or galaxies, one must determine the center of mass, establish a fixed reference frame with respect to this center and then employ the Newtonian conceptions of time and space.

   Take the interaction of two galaxies, which is bound to raise the most immediate objections.

   These two objects are moving with respect to each other and due to the influence of the one on the other by the equal and opposite forces of action and reaction. It does not matter how far apart they might be. Put your clock at the center of mass and measure!

   We have touched upon why current ideas of time and space are now such a clusterfuck in several previous books, but not in a rigorous and exhaustive way. Currently, we have bigger battles to fight and fish to fry! In addition, it is hard to know quite where to start to correct misconceptions that are so obviously wrong. Unfortunately, people currently seem quite proud of the fact that physics is crazy and mind boggling.   Mind boggling!   Oh my gosh!

        Coming soon: "On Time and Space" and "On TIme and Temperature".

Interlude: More Epigraphs !

    Thus energy, like the atom, is more and more divested of all sensuous meaning with the advance of knowledge. This development appears most clearly in the concept of potential energy, which even in its general name points to a peculiar logical problem. As Heinrich Hertz has emphasized, there is a peculiar difficulty in the assumption, that the alleged substantial energy should exist in such diverse forms of existence as the kinetic and the potential form. Potential energy, as it is ordinarily conceived, contradicts every definition that ascribes to it the properties of a substance; for the quantity of a substance must necessarily be a positive magnitude, while the totality of potential energy in a system is under some circumstances to be expressed by a negative value. Such a relation can, in fact, according to Gauss' theory of negatives, only be explained where that which is counted has an opposite, *i.e*, "where not substances (objects conceivable in themselves) but relations between two objects are counted."

⇔

    Thus the passive *fixity*, established by science at certain points, is an element in its own activity. In fact, it is justified and unavoidable, that science should condense a wealth of empirical relations into a single expression, into the assumption of a particular thing-like "bearer." The critical self-characterization of thought, however, must analyze this product once more into its particular factors, although it conceives this product as *necessary* for certain purposes of knowledge. This is done because critical thought is not directed forwards on the gaining of new objective experiences, but backwards on the origin and foundations of knowledge. The two tendencies of thought here referred to can never be directly united; the conditions of scientific *production* are different from those of critical *reflection*. We cannot use functions in the construction of empirical reality and at the same time consider and describe them.

⇔

    The goal of theoretical physics is and remains the universal laws of process.

--- Ernst Cassirer
Substance and Function

Math and physics:

The reason mathematics describes physics so well is because the two were devised in tandem, *and* because physical processes are mechanistic and deterministic.

Now, when physics tries to import independent mathematical ideas such as group theory or non-Euclidean geometry, the result is an unmitigated disaster. There are no 'internal symmetries' dictating the nature of physical processes, and for all intents and purposes space is Euclidean and independent of time.

Even well motivated and useful mathematical abstractions can lead to disaster when people begin to take them too seriously. The perfect example is the concept of the field and its antecedent concept; potential energy. As we have shown, potential energy is a useful placeholder for many mathematical operations but is not a real thing that inheres in particles, or a system of particles, or in fields.

The concepts of the field and potential energy originate in Classical physics, of course. Physicists have 'forgotten' that fields are a mathematical abstraction, or calculational convenience, *derived from the equal and opposite forces observed between two isolated interacting objects*. This abstraction is useful when one wants to study special, idealized situations. However, the idea of taking the limit of the force per unit charge as the value of a 'test charge' goes to zero is mathematically problematic, even in the Classical formulation. I don't think this problem has gone away.

Let's take classical electromagnetic waves as an example. A plane electromagnetic wave is not a series of sinusoidally oscillating electric and magnetic fields of fixed frequency. In our formulation, such a wave is a series of wave fronts consisting of photons. The photons themselves oscillate at the same frequency that determines their rate of emission from a given source.

The description or reduction of these photons in terms of electromagnetic waves is analogous to the characterization of the microscopic properties of an ideal gas in terms of the macroscopic variables of volume, temperature and pressure.

Now the universe is rife and lousy with fields of every description.
New problem? New field!

The main sin and major strike against General Relativity is that it is a field theory.
In addition, it has an 'infinite' number of predictions, many unphysical.
Falsification defined.

Physics and metaphysics:

'Modern' physics is replete with metaphysical assumptions whether the current devotees, or practitioners (if we are being gracious), care to admit it or not. In truth, one cannot call such a hodge podge of kludges and ad hoc assumptions metaphysics proper, as they reflect no underlying unifying or consistent philosophy.

Disembodied fields, collapsing wave functions, the vacuum, broken internal symmetries, particles traveling backward in time, etc., are not only silly ideas, but they can never be observed or empirically verified. The main argument for their reality is they make the math work out right. Given the number of tunable knobs in the standard model, plus renormalization, it is no wonder that people can make so many things work out to their satisfaction.

Currently, our new model is based on one simple *metaphysic*; particles 'want' to be at rest or achieve the lowest possible kinetic energy given the local environment. (It is still hard to express such ideas without using anthropological terms!)

All particles are connected by virtual photons; one virtual photon per particle pair. These photons are unbreachable and unbreakable. Each particle in the universe sees every other particle as a possible sink into which to dump their excess energy and momentum. This ongoing contest of mutual exchange causes particles to accelerate and gravitate towards or away from one another depending on their individual charges and spins.

This minimization process is expressed mathematically in equation (i) which we will reproduce here

$$\delta \int (m(t) - m_0)\, dt = 0 \qquad (5)$$

Photons and leptons:

In our model photons have mass as indicated in equation (j), which we reproduce here

$$m = h\upsilon/c^2 = \hbar\omega/c^2 \qquad (6)$$

Of course, the photon also has a momentum which we will write as

$$p = \hbar\omega/c \qquad (7)$$

The work done on a photon can be expressed as follows (see "On Rotation")

$$\int F\, dt = \int (dp/dt)\, dt = \int (\hbar/c)(d\omega/dt)\, dt \qquad (8)$$

In our model the universe is not expanding, and the shift in frequency of light from a distant galaxy should be derived from equation (8) where the force is due to the mass of the emitting galaxy, formulated from Newton's universal law in the usual way. There will be an additional complication due to the fact that the separation between the galaxy and the photon will be time dependent; $R = R(t)$. We shan't explore this further right now!

As demonstrated/hypothesised in "On Rotation" the **photon** also has a **magnetic moment!**

$$\mathbf{m}_\gamma = h/c\, \mathbf{i} \quad ; \quad \mathbf{i} = \mathbf{v}/|\mathbf{v}| \quad ; \quad v = c \qquad (9)$$

[equations (8) and (9) have been corrected] and, finally, the photon has a moment of inertia

$$I_\gamma = (\hbar\omega/c^2)(\lambda^2/4\pi^2) \qquad (10)$$

In our model the photon has an 'effective radius' equivalent to its wavelength, λ, as imagined in Figure C. We believe that this is how the photon is able to go through two slits of a similar separation "at the same time."

The **neutrino** also has a **magnetic moment**. This has been derived by an analogy with the magnetic moment of the electron (over the course of our investigations) and is given in equation (k). This magnetic moment is solely due to and is implicitly dependent on the spinning mass.

Accordingly, we find that the **electron magnetic moment** must have an additional magnetic moment term due to the mass contribution

$$\mu_e = (e/m_e + 1)(\hbar/2c)(1 + \tfrac{1}{2} v^2/c^2 + \tfrac{3}{8} v^4/c^4 + \ldots) \qquad (11)$$

Note that the magnetic moment is velocity dependent. If we neglect the contribution from the mass term for now then equation (11) reduces to

$$\mu_e = (e\hbar/2m_e c)(1 + \tfrac{1}{2} v^2/c^2 + \tfrac{3}{8} v^4/c^4 + \ldots) \qquad (12)$$

and we can see in our model the 'anomalous' correction of the standard model is cast in terms of an expansion in $(v/c)^2$ rather than α. Remembering that one of the definitions of α is in terms of the velocity of the electron in the ground state of the hydrogen atom, it seems these two formulations could be reconciled. We shall not pursue this further right now. We also note that in our model the running of α is also expressed in an expansion of $(v/c)^2$.

$$\alpha = \alpha_0 (1 + (v/c)^2 + (v/c)^4 + \ldots) \qquad (13)$$

$$\alpha_0 = e^2/4\pi\varepsilon \hbar c \qquad (14)$$

Whether this is "covariant" or whether it matters we shall also not consider right now.

We remind the reader that in our model the factor of e/m_e is dimensionless so the magnetic moment of the electron (and the photon and the neutrino) has units of length • mass and can be defined as "the first moment of mass." We have defined the 'radius' of the electron at rest to be equivalent to the Compton wavelength, so, we can write for the first moment of mass

$$\lambda_e m_e = (h/m_e c) m_e = h/c \qquad (15)$$

In this simple hand waving argument, at least the units come out right!

Finally, we note that in our model the second moment of mass, or the moment of inertia, is

$$I = m\lambda^2/(2\pi)^2 \qquad (16)$$

This formula applies for the electron and the neutrino and the photon.

The weak and strong forces:

While we are on the subject of the running of the magnetic moment and the electromagnetic coupling constant it seems a good time to (re)introduce the reader to our derivation of the coupling constants for the other three forces.

The gravitational coupling constant is

$$\alpha_G = (m_e^2 G)/(\hbar c) \, (1 + (v/c)^2 + (v/c)^4 + \ldots) \tag{17}$$

The weak coupling constant is

$$\alpha_W = (m_\nu^2 G)/(\hbar c) \, (1 + (v/c)^2 + (v/c)^4 + \ldots) \tag{18}$$

And finally, the strong coupling constant (14) is

$$\alpha_S = (G/4\pi\varepsilon)^{1/2} \, (2 m_e e/\hbar c) \, (1 + \tfrac{1}{2} v^2/c^2 + \tfrac{3}{8} v^4/c^4 + \ldots) \tag{19}$$

In the universal model, all the 'constants' run, because *the fundamental coupling charge of a particle is the relativistic mass-energy* of the particle. There is no weak charge or color charge in our model and the weak and strong forces are simply the manifestation of gravity and electromagnetism at subatomic scales and relativistic velocities. Obviously, the weak force is simply the gravitational force and is weak due to the mass of the neutrino!

There is no need for renormalization.

In our model the recently discovered Higgs boson is the gravitationally bound state of a neutrino and an antineutrino. Each neutrino can radiate a ('virtual') photon, each of which produces a lepton antilepton pair, or the neutrino and antineutrino can 'annihilate' to produce a virtual photon then producing a lepton antilepton pair leading to the following decay modes;

$$H \to \nu\bar\nu + l^+l^- + l^+l^- \, ; \, l = e, \mu, \tau \tag{20}$$

$$H \to l^+l^- \, ; \, l = e, \mu, \tau \tag{21}$$

In equation (21) the virtual photon can also produce a photon anti-photon pair!

With the proper choice of radius, using the known mass of the neutrino, and applying the uncertainty principle, it should be easy to estimate the mass of our Higgs. We leave this as an exercise for the reader!

The propagator:

In our model the mass dependence, and short range, of the 'weak' force is due to the tiny mass of the neutrino in the coupling parameter (plus the factor of G) as shown in equation (18) and is *not* due to a massive exchange boson as in the standard model. The only exchange boson is the virtual photon. Technically, the virtual photon is not exchanged, but represents a constant tethering of two particles, allowing for a continuous exchange of energy and momentum consisting of 'equal and opposite action and reaction'.

The idea of a constant and continuous exchange of energy and momentum between two interacting particles is best illustrated in the construction of the propagator between two 'spinless' leptons in "Quarks and Leptons" (Halzen and Martin) where it is simply explained that the *current of either particle* can be chosen as providing an electromagnetic field influencing the other particle and *the resulting photon propagator between the two is the same*. We take this most sensible and excellent presentation from pages 84-87 and refer the reader there for all the details. Their derivation is in terms of the 'lowest order' approximation using field theory, although we shall adapt it to motivate our field free model.

In our reformulation, A^μ represents the 'wave function of the virtual photon' and not the electromagnetic potential. Hence, we needn't disregard the higher order contribution from the e^2A^2 term because it no longer is relevant. The transition amplitude is

$$T_{fi} = -i \int j_\mu{}^1 A^\mu d^4x \quad ; \quad A^\mu = -(1/q^2) j^\mu{}_2$$

and the invariant amplitude is

$$M \sim e^2(-i\, g_{\mu\nu}/q^2) \quad ; \quad \text{where } -i\, g_{\mu\nu}/q^2 \text{ is the photon propagator.}$$

This is all very rough of course and we may have not needed to include the invariant amplitude to illustrate the differences between the standard approach and our new model, but it conveniently introduces the factor of e^2 into the discussion.

Now, besides the fact that our model is 'linear' in A^μ, the other important point to emphasize is in our model *the electric charge is mass dependent* and we must replace e with $m(e/m_e)$, where m is the relativistic mass. Hence, higher order contributions to the matrix element will not be due to or derived from a perturbative expansion in α *per se*, but will arise from the increase in the masses of the particles in high energy interactions. See equation (13).

The Dirac equation:

As we mentioned in the section on Symmetry, we have *successfully removed* the rest mass term from the traditional Dirac equation and placed it in the wave function instead. We believe this has several important consequences. Firstly, it should make the construction of the Dirac Lagrangian easier and more natural. Secondly, it may remove peoples' belief that having removed the rest mass term from the standard model Dirac Lagrangian 'by hand', that there is a need to reintroduce or reinstate the particle mass somehow by recourse to the Higgs field.

We're not sure why people believe crazy things, to be honest !

So the universal Dirac equation is

$$H \psi = c \alpha \cdot p \, \psi \qquad (22)$$

$$i\hbar \, \partial \psi / \partial t = -ic\hbar \, \alpha \cdot \nabla \psi \qquad (23)$$

and is satisfied by, or solved with, the universal wave function

$$\psi = \exp(i(p \cdot x - (m-m_0)c^2 t)/\hbar) \qquad (24)$$

$$mc^2 \psi = c \, \alpha \cdot p \, \psi + m_e c^2 \psi \qquad (25)$$

The free particle, spin up, solution of the Dirac equation for an electron (positive helicity) is

$$\psi = \begin{vmatrix} 1 \\ 0 \\ \sigma \cdot p/mc \begin{vmatrix} 1 \\ 0 \end{vmatrix} \end{vmatrix} \exp(i(p \cdot x - (m-m_0)c^2 t)/\hbar) \qquad (26)$$

$$\sigma \cdot p/mc = \begin{vmatrix} p_z & p_x - ip_y \\ p_x + ip_y & -p_z \end{vmatrix} \qquad (27)$$

→ the determinant = $-(p_z^2 + p_y^2 + p_x^2)$ → $-(p_z^2 + p_y^2 + p_x^2)/mc = -1$ (28)

→ the square = $(p_z^2 + p_y^2 + p_x^2)\mathbf{1}$; **1** = 2x2 unit matrix (29)

The matrix $\sigma \cdot p/mc$ *is the factor* that turns our ordinary column vector *into a spinor*, which is technically classified as an Euclidean tensor. So the universal model involves scalars, vectors, and tensors: all Euclidean.

A spinor is a vector with zero 'length' which means that two of the three components must be imaginary. In our example, only p_z is real; the direction of propagation!

Since the spinor precesses about the direction of propagation, the average values of p_x and p_y are zero, but this does not mean they are undetermined or do not have distinct values and directions at any given time.

The form of equation (21) turns up again and again; i.e. rotations, angular momentum ... however, equation (21) is not the result of a rotation or 'cross product', as in the other examples.

We hope to explore the role and significance of complex numbers in modern physics more in our next book.

The Dirac equation with the universal interaction term added is

$$i\hbar \partial \psi / \partial t = -ic\hbar \alpha \cdot \nabla \psi + mc^2 A^\mu \qquad (30)$$

where A^μ represents the wave function of the virtual photon mediating the interaction with a second particle. <u>It does not represent a potential field.</u> For a single particle interacting with several other particles, A^μ would be represented by a Fourier sum or integral over all the relevant virtual photon frequencies.

no more fields

[This section has been cut and pasted from "Toward a Metaphysics of Mass and Motion"]

[Sorry for the Clip Show!]

Postmortem physics:

Let us be brief.

Fields are not real. The quantum mechanical vacuum is not real. Potentials are not real. Gauge symmetry is not real. Potential energy is not real. The Higgs boson is not real.

Wave functions are not real.

None of these things are real.

Mass is real. Spin is real. Kinetic energy is real. Acceleration is real.

Photons are real. Electrons are real. Neutrinos are real. They are all mass in motion.

There is only mass in motion.

We can't even, as they say.

: /

Conclusion:

This book did not turn out quite as expected, but then they never really do. We haven't calculated much of anything nor introduced too many new ideas. The Postmortem was not the vivid vivisection or blistering rebuttal of the standard model that we had envisioned. I become weary just thinking about it. I weary of wondering why people think it is so wonderful. I say, let the books speak for themselves.

I say, let people think for themselves. It is possible! I must believe it.

So, let us consider this book an intermezzo.

In our next offering(s) we expect to take a closer look at spin, the Born rule, the Born approximation, the uncertainty principle, the Dirac equation, and the Hydrogen atom, etc. I don't expect we will still actually calculate much of anything, but I do expect we will finally finish laying a firm foundation for our new 'universal model' and provide the tools for others to do useful work.

Books by Greg Feild:

1. "A quantum mechanical theory of gravitational interactions"
   CreateSpace Independent Publishing, 8/29/2016

2. "Observations on the quantum mechanical nature of gravity"
   CreateSpace Independent Publishing, 10/8/2016

3. "On gravitation and electric charge"
   CreateSpace Independent Publishing, 10/29/2016

4. "On spin, mass, and charge"
   CreateSpace Independent Publishing, 11/29/2016

5. "On angular momentum, acceleration, and absolute motion"
   CreateSpace Independent Publishing, 1/1/2017

6. "The Sinister Universe"
   CreateSpace Independent Publishing, 3/1/2017

7. "On Parity and Isospin"
   CreateSpace Independent Publishing, 4/11/2017

8. "Reflections on the Sinister Universe"
   CreateSpace Independent Publishing, 5/12/2017

9. "On Current Physics"
   CreateSpace Independent Publishing, 6/11/2017

10. "A Critical Examination of Classical and Quantum Mechanical Waves"
    CreateSpace Independent Publishing, 6/18/2017

11. "On wave particle duality and the quantum of action"
    CreateSpace Independent Publishing, 7/6/2017

12. "On matter, mass, and motion"
    CreateSpace Independent Publishing, 9/14/2017

13. "On action and reaction"
    CreateSpace Independent Publishing, 9/24/2017

14. "A quantum mechanical theory of everything"

CreateSpace Independent Publishing, 11/5/2017
15. "On Interaction"
    CreateSpace Independent Publishing, 4/21/2018

16. "On Rotation"
    CreateSpace Independent Publishing 8/19/2018

17. "Revenge of the Sinister Universe: The Reality of Everything'
    CreateSpace Independent Publishing, 9/4/2018

18. "On Math, Physics, and Metaphysics"
    CreateSpace Independent Publishing, 10/1/2018

19. "On Quantum Mechanics"
    CreateSpace Independent Publishing, 10/15/2018

20. "On Epistemology and Ontology"
    CreateSpace Independent Publishing, 10/21/2018

21. "Toward a Metaphysics of Mass and Motion"
    CreateSpace Publishing, 10/29/2018

Compilations:

A. "The Universal Model of Our Sinister Universe: The First Ten Books"
   CreateSpace Independent Publishing, 7/2/2017

B. "The Canons of the Sinister Universe:
   The Last Four Books on the Universal Model of Our World"
   CreateSpace Independent Publishing, 11/5/2017

C. "The Return of the Sinister Universe: The Immaculate Collection"
   CreateSpace Independent Publishing, 9/4/2018

D. "The Battle for the Sinister Universe: The Heuristics"
   CreateSpace Independent Publishing, 10/29/2018

Resources:

Quantum Field Theory
Claude Itzykson, Jean-Bernard Zuber

Atomic and Quantum Physics
H. Haken, H.C. Wolf

Modern Elementary Particle Physics
Gordon Kane

Classical Dynamics of Particles and Systems
Jerry B. Marion

Foundations of Electromagnetic Theory
John R. Reitz, Frederick J. Milford, Robert W. Christy

Quantum Physics
Rolf G. Winter

Gauge Theories in Particle Physics
I. J. R. Aitchison and A. J. G. Hey

Quarks and Leptons: An Introductory Course in Modern Particle Physics
Francis Halzen, Alan D. Martin

Quantum Field Theory
F. Mandl, G. Shaw

Theoretical Mechanics of Particles and Continua
Alexander L. Fetter, John Dirk Walecka

The Theory of Spinors
Elie Cartan

Elementary Modern Physics
Richard T. Weidner, Robert L. Sells

Quantum Mechanics
Claude Cohen-Tannoudji, Bernard Diu, Franck Laloe

The misanthropic principle:

everyone is the worst !

News for parrots:

no parrots were involved.

www.ingramcontent.com/pod-product-compliance
Lightning Source LLC
Chambersburg PA
CBHW080817220526
45466CB00011BB/3590